DIRECTOR'S CHOICE
THE ROYAL INSTITUTION

Katherine Mathieson

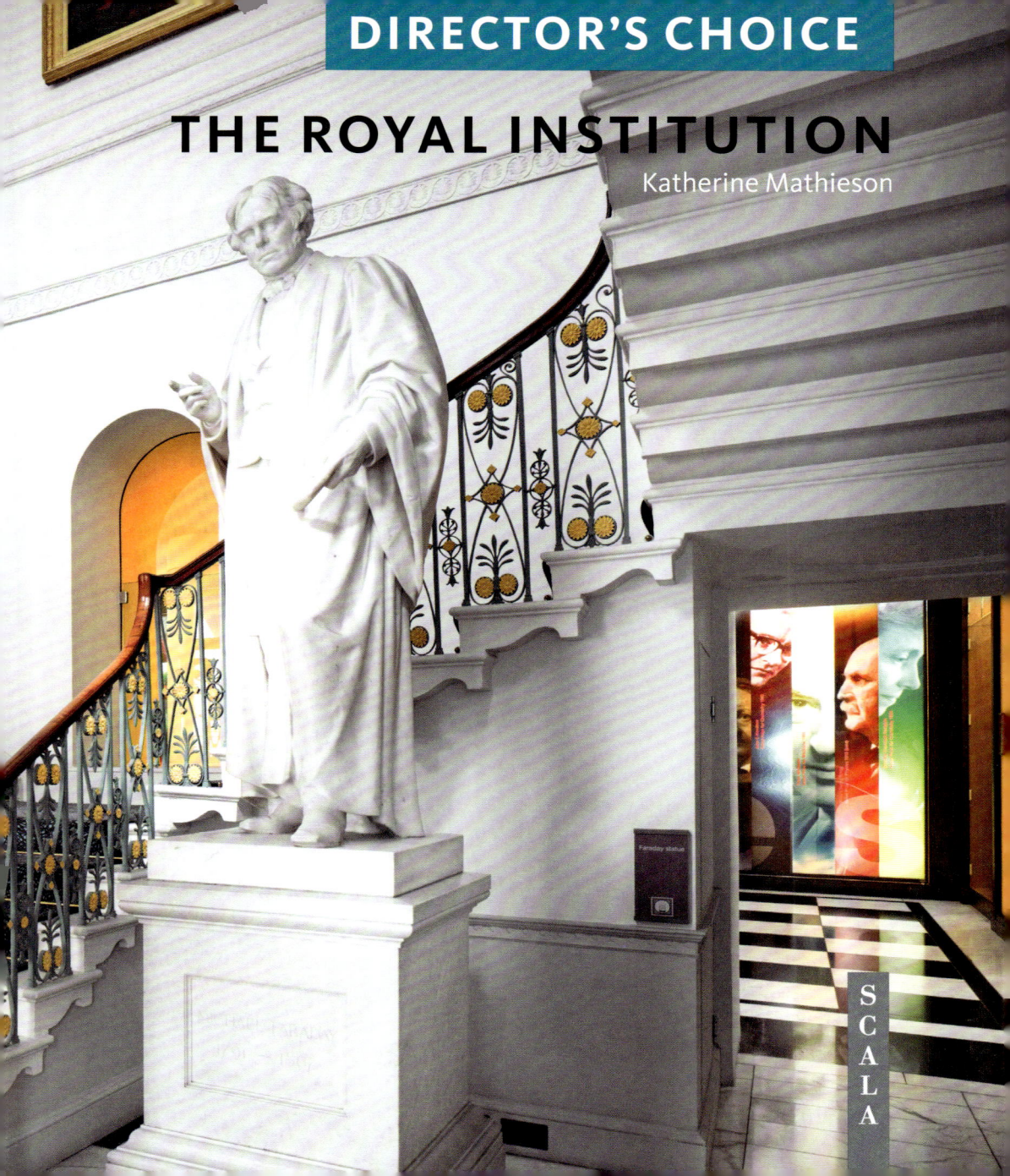

DIRECTOR'S CHOICE

THE ROYAL INSTITUTION

Katherine Mathieson

SCALA

INTRODUCTION

THE ROYAL INSTITUTION (RI) of Great Britain is a unique and special place. Founded in 1799 as a venue for communication and discussion to 'diffuse the knowledge and facilitate the general introduction of useful mechanical inventions and improvements ... and to teach by courses of philosophical lectures and experiments the application of science to the common purposes of life', the Ri has always been driven by the idea that science should be visible and useful. Occupying the same building in Albemarle Street, London, since its founding, the Ri has adapted it to create laboratories and spaces for discussions, lectures and demonstrations.

Fundamental to the Ri's existence are its internationally significant collections, which include scientific apparatus and objects, archival material, photographs, film and rare books. The majority are items made or used in the building (or indeed, are parts of the building itself), or belonged to prominent Ri members or staff, illustrating the research and scientific communication that has occurred at the Ri representing the uses and actions of the premises.

Each artefact, whether apparatus, archival manuscript or piece of artwork, has been kept not only to remember a scientific development but to continue to communicate past discoveries engaging with the public. Most have never left the Ri, apart from for conservation, or for protection during the Second World War.

Together, the collections reflect many of the main advances in the understanding and application of physics, chemistry, engineering and materials from the last 200 years. The collections help ground us and remind us that science is ongoing and not the work of one individual but often collaborative and reliant on the efforts of those that have come before.

In journeying through the collections, a common theme emerges: illustration. Many of the items – whether a piece of apparatus, a notebook, painting or something else – were developed to make the unknown visible. Unseen, unfelt phenomena became tangible. The experimental apparatus that led to a discovery was reused in future lectures to show the same discovery to the wider world. The research notebooks are often heavily illustrated with descriptions and drawings, conveying both the inspiration for and the conclusion of a fleeting experiment so that it lives on. Photographs and paintings help to illustrate not only a moment in time, but also the individuals who participated.

Throughout its history, the Ri has been an inclusive and welcoming place; one of the Institution's most iconic residents, Michael Faraday, started as an apprentice

This watercolour on paper, by Harriet Moore (*c*.1850s), depicts part of the original open laboratory in the lower ground floor of the Ri. See pp. 28–29.

bookbinder and lab assistant before going on to become one of the most brilliant experimental scientists of his time. The audiences at Ri lectures have always had a mixture of backgrounds and genders; women were welcomed as members and attendees from the beginning.

To this day, the Ri continues with this endeavour, being a place of discourse, dialogue and demonstration, enabling people from all backgrounds, and whatever their level of science knowledge, to come together to share, understand and communicate the science that underpins our modern world.

Left: Alexander Blaikley's lithograph (1855) shows Michael Faraday giving a Christmas Lecture at the Royal Institution, 1855 (see pp. 26–27).

Above: Royal Institution Lecture Theatre, *c.* 2020 (see pp. 8–9).

Lecture Theatre of the Royal Institution, constructed 1801
21 Albemarle Street

THE FOUNDING AIMS OF THE Ri in March 1799 were twofold: 'the speedy and general diffusion of the knowledge of the new and useful improvements, in whatever quarter of the world they may originate'; and 'teaching the application of scientific discoveries to the improvement of arts and manufactures in this country and to the increase of domestic comfort and convenience'. To fulfil these aims the Ri needed a permanent home suitable for the communication and teaching of scientific endeavours, discoveries and innovations.

The construction of the lecture room, or Theatre as it is known today, began in 1799 under the direction of Thomas Webster (1773–1844). Webster's design drew on the steeply raked bench seating of an anatomy theatre, allowing every audience member a clear view of the centre stage. Webster added a gallery for those wishing to attend but be less observed. This space was for artisans, to segregate them from their employers who might be sitting below. A raisable dome in the ceiling brought light into the space and created excellent acoustics.

The Theatre remained much the same until 1927 when there was an electrical explosion in the basement. There was a realisation that it needed to be upgraded and modernised. Over a period of three years, when all Ri lectures were delivered from other locations, the gallery was shortened and columns removed, fire exits were added, and the benches replaced with seats. These changes reduced the capacity from nearly 1,000 to 440.

The Lecture Theatre still sits at the heart of the Ri today, hosting our thriving lecture and events programme, and the renowned Christmas Lectures for children. It is a space that still inspires not only those lecturing but also those in attendance, evoking the history of all those who have used the space over the last 200-plus years and continuing to play its part in the communication of science.

Michael Faraday (1791–1867)

Ten of Michael Faraday's scientific notebooks, 1820–62

Variable dimensions
Bequest of Michael Faraday in 1855 to be given to the Royal Institution on his death
RI MS F/2/A-I.

These ten scientific notebooks recorded Michael Faraday's experimental research into various subjects such as the liquefaction of gases and electromagnetic rotations as well as investigations of substances sent to him. In Faraday's original order, the notes are roughly chronological but with exceptions.

Faraday's origins were humble. The son of a blacksmith, he grew up in Newington Butts (now Elephant and Castle). On leaving school at 13, he undertook an apprenticeship as a bookbinder. A random set of events led to him becoming a laboratory assistant from 1813, the start of his career at the Ri.

These notebooks form part of the Ri's archive, specifically part of the Faraday Collection. Within these notebooks Faraday documented some of the most important discoveries of the nineteenth century. They include his work on electromagnetism, lines of force and chemical investigations. His meticulous notes provide clear evidence of his processes of experimentation and scientific research, accompanied by detailed descriptions and drawings. The Ri holds a unique archival collection that documents the development of the Institution as well as the scientific investigations of the professors or visiting lecturers and their communications with each other and with wider society. At the centre of this archive is the material produced by Faraday.

True to the nature of the Ri, Faraday was a great advocate for promoting science to as broad an audience as possible. It is therefore appropriate that in 2016 these ten notebooks were placed on the UNESCO Memory of the World Register – UK Entry, becoming part of the Register's affirmation that the world's documentary heritage belongs to everyone, that it should be preserved, protected, and made permanently accessible to all.

Michael Faraday (1791–1867)

Cylinder Electrostatic Generator, c.1806

Experimental Apparatus, glass and wood, 320 mm × 365mm × 160 mm
Donated by James South, to whom Faraday gave the apparatus.
RIAC 00004

This object is a hand-constructed electrostatic generator, representing Michael Faraday's first ventures into experimental science. It was made by him when he was around 15 years old.

Having left school at 13, Faraday initially worked as a paper boy before becoming an apprentice bookbinder at the age of 14 at the shop of George Riebau in Blandford Street, London, clearly demonstrating skill in writing and reading. Faraday liked to read the scientific books as he bound them, furthering his knowledge and continuing his education.

In 1806 Faraday tried to carry out his own experiments. Taking an old glass bottle, he placed it together with some wooden supports (now identified to be from pieces of bookbinding equipment) to see if he could make an electrostatic generator. Originally this machine would have had a small leather-covered block attached to the base, which allows static electricity to be generated between the glass and the leather.

While Faraday did not invent the electrostatic generator, he desired to build equipment that could illustrate easily the invisible, becoming a master of visualising scientific developments through description, drawings, apparatus, and demonstration.

Alessandro Volta (1745–1827)

Voltaic Pile, c.1800–06

Copper, zinc and blotting paper, 232 mm × 102 mm × 102 mm
Donated by Benjamin Leigh Smith, 1902
RIAC 00008

This object is an example of a very early battery, known as a Voltaic Pile, named after its inventor Alessandro Volta (1745–1827). Volta developed the battery in 1797 in Italy. It consists of a stack of copper and zinc discs sandwiched together with circles of blotting paper. The paper would have been soaked in an acid, which reacted with the metal around it, creating an electric current between the top and bottom discs.

The development of the battery helped to significantly increase scientific investigation, particularly in the area of chemistry, where batteries allowed for electrochemical experimentation, assisting in the isolation of elements throughout the early 1800s.

This particular battery has been in the collection of the Ri since the early twentieth century. It was personally given by Volta to Michael Faraday when he and Humphry Davy were on a tour of Europe from 1813 to 1814. Both Davy and Faraday were keen to meet Volta, who was then nearly 70, to acknowledge the importance of his invention of the battery, and finally had the opportunity while they were in Milan in June 1814.

Today, we take the easy availability of portable electric power for granted, from the phones in our pockets to the electric cars on our roads. This technology is still improving, with ongoing research into developing ever more efficient, green, and long-lasting batteries to power our everyday lives. The work of Alessandro Volta was a vital initial first step in this development.

Michael Faraday (1791–1867)

Earliest surviving electric motor, 1822
Also known as Faraday's mercury bath or electromagnetic rotation apparatus

Glass, copper alloy, iron, wood and shellac, 247 mm × 173 mm × 121 mm
Originally constructed and used at the Royal Institution
RIAC 00012

The simplicity of this object masks its significance as the earliest surviving electric motor. It was originally developed by Michael Faraday in 1821, building on the work of Hans Christian Ørsted (1777–1851) and André-Marie Ampère (1775–1836).

In 1820 Ørsted discovered that when an electric current flowed through a wire a magnetic field was produced around the wire. Ampère followed this by showing that the magnetic field was a circular one, effectively creating a cylinder of magnetism around it. Faraday was the first to understand the implication of these two discoveries, which revealed that if a magnetic pole could be isolated then it should move constantly around a current-carrying wire in a circle.

This apparatus, the only original surviving example made by Faraday, in 1822, consists of two glass cups resting on a wooden frame with a piece of thin wire suspended from the frame above, over the larger cup. The other side of the frame is bent down so that the end is suspended within the smaller cup.

In the original version, only the larger glass cup existed, and a bar magnet would have been secured within the base of the cup. If you filled the cup with mercury, an excellent conductor of electricity, and connected a battery to the apparatus, electricity would be sent through the suspended wire, thereby creating a magnetic field around it. This field interacted with the field around the magnet and caused the wire to rotate clockwise, transforming the electrical energy into mechanical movement for the first time.

This improved version of the apparatus, with two cups and one fixed wire and one static, could demonstrate both the wire rotating round a magnet and the mercury moving around the static wire at the same time. While demonstrating a new scientific principle, it also signifies a turning point in Faraday's contemplation of the world and the nature of electricity.

Humphry Davy (1778–1829)

Pure Element Samples of Sodium, Calcium, Magnesium and Chlorine,
1807–8

Sodium 175 mm × 175 mm; magnesium 160 mm × 165 mm; chlorine 141 mm × 62 mm; calcium 122 mm × 76 mm
Made and used at the Royal Institution
RIAC 00641-00644

Sir Humphry Davy was professor of chemistry at the Ri from 1802–23. He had made a name for himself working in Bristol at the Numatics Institute where he discovered nitrous oxide, a substance more commonly known today as laughing gas.

His enthusiasm for communicating his discovery brought him to the Ri to lecture and he was promptly offered a role in the Institution, allowing him use of the extensive resources and laboratories.

Davy's research relied heavily on electricity, aided in particular by the development of the battery. He made great advances in the field of electrolysis, using the power of electricity to help understand the building blocks of life.

At the start of the 1800s only 26 elements were known to exist. Davy realised it was possible that elements were combined in chemical compounds and started to study ways of breaking these compounds apart. Using electricity, he passed a current through the substances or compounds to isolate the elements.

Davy initially isolated sodium in 1807 and went on to either prove the existence of or name a further eight chemical elements, including potassium (1807), chlorine (1808), magnesium (1808), strontium (1808), calcium (1808), barium (1808), boron (1808) and iodine (1811). On display at the Ri are these small glass bottles containing Davy's original demonstration samples of four of these elements: sodium, calcium, magnesium and chlorine.

PURE ELEMENT SAMPLES OF SODIUM, CALCIUM, MAGNESIUM AND CHLORINE

FOUR LECTURES
being part of a Course on
The Elements of
CHEMICAL PHILOSOPHY
Delivered by
SIR H. DAVY
LLD. Sec RS. FRSE. MRIA. MRI. &c &c.
AT THE
Royal Institution
And taken off from Notes
BY
M. FARADAY
1812

Michael Faraday (1791–1867)

Faraday's Notes on Davy's Lectures, 1812

Paper, leather, board and iron gall ink, 210 mm × 175 mm × 40 mm
Bound manuscript donated by John Davy, brother of Sir Humphry Davy
RI MS F/4/A

This manuscript was created by Michael Faraday in 1812. At the time, he was a newly qualified bookbinder, showing a passion for science and learning. He was offered tickets to attend Sir Humphry Davy's lecture series on 'The Elements of Chemical Philosophy' at the Ri, featuring information on radiant matter, chlorine, simple inflammables and metals, and he jumped at the chance.

Faraday took detailed notes of the lectures, adding small illustrations and a thorough index. He bound his notes using his bookbinding skills and resources available to him, and asked to meet with Davy at the Ri, where he offered the book as a gift while also asking for a job. Faraday was eventually given a role assisting Davy and the rest, as they say, is history.

This treasured volume is especially important because it uniquely links the Ri's archive, object collections and history with the building itself. On page 167, Faraday recounts that Davy 'exhibited a specimen of Chlorine gas'. This specimen remains within the collections of the Ri, and the Lecture Theatre where this demonstration took place and where Faraday came to listen and take notes still stands.

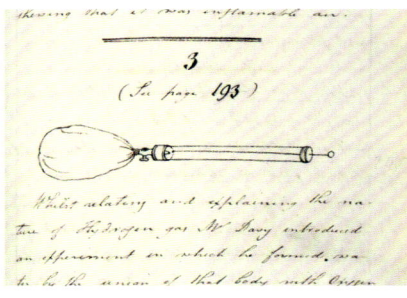
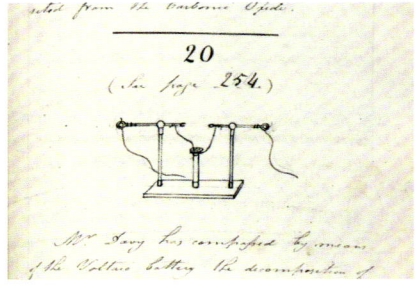

Thomas Hosmer Shepherd (1793–1864)
Façade of the Royal Institution, c.1838

Watercolour on paper, 363 mm × 437 mm
Potentially commissioned by the Royal Institution in c.1838
RIIC 0277

This watercolour, painted by Thomas Hosmer Shepherd in around 1838, shows the colonnaded façade of the Royal Institution in its home on Albemarle Street, Mayfair.

The Ri moved into the Street in 1799, having purchased three adjacent buildings with the funds from the initial proprietary membership. The Ri started life in this location as it was a short distance from the Royal Society of London, then in Burlington House, and near to the Royal Court at St James's Palace – thus placing it very much in the heart of court life. While construction took place on the inside to unify the spaces, the outside remained a collection of disjointed frontages.

Under Michael Faraday's direction in 1837 there was a strong motivation to construct a façade worthy of the Institution's scientific reputation and to show that the building was something different from the hotels, clubs, houses and publishing companies that lined the rest of the street.

The iconic façade, designed in 1837 and which is still in place today, was devised by Lewis Vulliamy (1791–1871) and was based on the Tempio di Antonino. Faraday had visited the ancient temple when he was in Rome with Humphry Davy in 1814.

John Tyndall (1820–1893)

Blue Sky Tube and Stand, c.1869

Glass and copper alloy, 495 mm × 888 mm × 109 mm
Made and used at the Royal Institution
RIAC 00330

In 1869 Professor John Tyndall was experimenting with gases in the basement of the Ri. He was investigating light and its properties. Using a glass tube with metal fittings at either end, Tyndall began looking into microscopic particles in the atmosphere.

In one experiment Tyndall shone a white light through this glass tube and slowly filled it with smoke. When he looked at the tube from the side, the light no longer appeared white but was instead blue. When he looked at the opposite end of the tube from the light source, the light appeared reddish.

Tyndall realised that the experiment showed how we see the colours in the sky, as a result of the Sun's light being scattered through particles in the upper atmosphere. As blue light has a shorter wavelength than red light it is more easily scattered, making the sky appear blue in our eyes. Several 'Blue Sky' tubes were made for Tyndall's experimentation, all by Harvey and Peake of London which expertly crafted the glass tube and copper-alloy fixings, gas tapes and stand.

This popular experiment is often reproduced in schools, and certainly within our education programmes at the Ri. Taking a fish tank of water and adding a little milk, you can shine a torch through the water from the outside of the tank and see the same effects as Tyndall did more than 150 years ago.

Alexander Blaikley (1816–1903)

A Christmas Lecture at the Royal Institution, 1855

Gouache over lithograph, 500 mm × 687 mm
Gift from Vernon Blaikley, 23 September 1931
RIIC 0006

This coloured lithograph by Alexander Blaikley shows Michael Faraday presenting the 1855 Christmas Lectures on 'The Distinctive Properties of Common Metals'. Within the image you can see a packed audience of adults and children occupying the Lecture Theatre of the Ri. Also visible is Prince Albert, the Prince of Wales (later King Edward VII), and Prince Alfred who does not appear to be paying much attention.

Today, the lithograph is displayed outside the Lecture Theatre, where the Christmas Lectures are still held, along with the original blocks of wood representing the different metals discussed by Faraday, which within the image are visible above the fireplace.

The Christmas Lectures were created by Faraday in 1825, for the purpose of lecturing on a specific subject to a young audience, originally those aged 15 to 20. The series has occurred every year since then, apart from during the Second World War, and is still the only Ri programme where an adult must be accompanied by a child and not the other, more usual, way round. The Christmas Lectures have been broadcast as full series on television since 1966.

During his lifetime, Faraday gave 19 series of Christmas Lectures from 1827 to 1860 and his original lecture notes still exist in the Ri archive. He often repeated the subject matter: for example, his most popular series on the 'Chemical History of a Candle' was delivered three times and was eventually published in book form in 1861.

A CHRISTMAS LECTURE AT THE ROYAL INSTITUTION

HARRIET MOORE (1801–1884)

*Rooms at the Royal Institution, c.*1850s

Watercolour on paper, various dimensions
Donated by various relations of Michael Faraday
RIIC 0282 & 0283, RIIC 0446-0449

THIS COLLECTION OF WATERCOLOURS, painted by Harriet Moore date from the 1850s and depict areas of the Royal Institution building, both public and private. Moore was given access to this world by her friendship with Michael Faraday and his wife Sarah. Through her illustrations we have fascinating evidence of how Faraday lived and worked.

Detailed observations of equipment and even experimentations can be seen in the watercolours of the large open laboratory space and the small Faraday Laboratory, both on the lower-ground floor of the Albemarle Street building. The large lab was dismantled in the 1870s, to make way for a more modern space and therefore the information that these paintings contain is vital for understanding how the space was laid out and used, as it was fundamental in Faraday's discoveries of electromagnetism and electrochemistry.

Faraday's Laboratory was painted by Moore in the early 1850s. It was in this space, which survives to this day, that Faraday showed in 1845 that light and magnetism were connected. In the same year he used a giant electromagnet, made of half an anchor ring that had its top cut off (seen underneath the table in the middle of the painting) to demonstrate that all matter had magnetic properties.

Mirror, 1802

Glass, wood and gilt, 700 mm diameter
Commissioned by the Royal Institution in 1802
RIOC 0062

This mirror was purchased by the Members of the Ri in 1802 at a cost of £20. It was originally in the first Lecture Theatre but was moved to the Library in November 1804 and has remained over the mantelpiece there ever since.

A circular, gilt, convex mirror, its outward-curving surface plays with the reflection. When you look into it, you see yourself looking back out, but expanding on its curve and the skill with which it was made, in a mirror such as this your reflection sometimes appears larger or smaller and occasionally upside down.

Having a convex mirror on display became quite fashionable in the eighteenth century, with gilded circular versions extremely popular in England by 1803. Playing a key role in the running of the Regency household, convex mirrors were often placed above the dining-room sideboard, allowing the butler to discreetly keep an eye on dinner guests while keeping his back turned.

The skill involved in creating this object means that we can regard it as an experiment in design and function, and not just a room adornment. The Ri's mirror is flawless and allows you to view the whole Library with little distortions to people's reflections.

It is possible that the mirror was moved to the Library so that staff could quietly observe all the Library users, monitoring how they viewed the books on the open shelving. The Library had a vast and rich collection and there is ample evidence in the archives of unwarranted removal and theft.

There are other convex mirrors in the collection, mainly Victorian copies, but none are made with such skill.

Section of Meteorite from Toluca, Mexico, discovered in 1776

Iron and nickel, 13 mm × 198 mm × 140 mm
Gift to the Royal Institution in c.1803
RIAC 00267

This small, shiny piece of metal is a section of a meteorite discovered in Toluca in Mexico in 1776.

While there are many examples of meteorites found on Earth, iron meteorites such as the one our section comes from are rare. Originally made up of large, scattered lumps with a combined weight of approximately 3 tonnes, the Toluca meteorite's composition is nearly 91 per cent iron (Fe) and just over 8 per cent nickel (Ni). It was discovered in the middle of a crater and is thought to have fallen to Earth more than 10,000 years ago. Our segment probably made its way into the possession of the Ri at a time when it was creating an extensive collection of mineralogical samples in the early 1800s.

Sadly, little of this collection remains, but we are lucky to have this sample as it is one of the few objects that visitors are allowed to handle.

While the collections of the Ri are expansive, covering a range of scientific developments and areas of research, most objects are fragile and have glass components. There is very little within the collection that can be handled by the public or non-museum professionals, and even less that children can touch. This meteorite fragment is an exception. Seeing adults and children alike holding something so old is lovely to see. It is an extraordinary experience to hold something that has been in the deepest areas of space and has existed for 4.5 billion years.

Michael Faraday (1791–1867)

Electromagnetic Induction Ring and Electrical Generator, 1831

Iron, copper wire, calico, leather (generator only), 13 mm × 198 mm × 140 mm
Induction Ring: 32mm × 170mm; generator: 280mm × 220mm × 40mm
Made and used at the Royal Institution
RIAC 00020, RIAC 00022

These two world-changing objects were made in the basement of the Ri in 1831 by Michael Faraday. Constructed from odds and ends, they have changed our world beyond all recognition and without them we would not be able to create electricity or control its power.

In late August 1831, Faraday wrapped copper wire bound in calico in two sections around a large iron ring. This work took approximately ten days and at the end of it, Faraday had created the world's first induction ring. This ring could change the voltage of an electric current, becoming the first ever transformer. Using this very ring, Faraday discovered electromagnetic induction, realising that induction produced a transient electrical current rather than a continuous one.

Just over a month later, Faraday wound a coil of insulated copper wire around a long iron bar or magnet. When the magnet was moved through the coil an electric current ran through the wire. By pushing and pulling the magnet through the coil, Faraday was able to generate electricity for the first time. The final generator was completed on 17 October 1831.

Today, variations of both these objects are used in every power station in the world, whether nuclear, gas, water or wind powered. It is incredible to think of the astounding impact these two curious devices have had on our lives.

ELECTROMAGNETIC INDUCTION RING AND ELECTRICAL GENERATOR | **35**

Henry Jamyn Brooks (1865–1925)

Sir James Dewar Lecturing at a Friday Evening Discourse, 1904

Oil on canvas, 1665 mm × 2741 mm × image
Gift from Ludwig Mond and Members of the Royal Institution, May 1906
RIIC 0050

This painting, by Henry Jamyn Brooks, depicts a Friday Evening Discourse given by Sir James Dewar (1842–1923) at the Royal Institution. The audience includes many famous people from science, industry and politics, such as Lord Rayleigh, Lord Kelvin and Prime Minister Arthur Balfour. This was the first time Dewar, Director of the Davy Faraday Research Laboratory (DFRL) at the Ri, liquefied hydrogen in public, by cooling it to -252.9°C and placing it within a vacuum flask to minimise heat loss and keep the gas liquid for as long as possible.

The Friday Evening Discourse Lecture Programme was started by Michael Faraday in 1825. It has run each year since, although on one Friday a month for nine months of the year rather than every Friday, until 2020/21 when the programme was temporarily suspended due to the COVID-19 pandemic. The programme was re-established in 2022.

Originally grand affairs only open to members and their guests, Discourse Lectures usually took place every Friday and sometimes Thursday. The series was specifically created for demonstrating cutting-edge science and scientific announcements, providing a stage for scientists to introduce their latest research, accompanied by numerous demonstrations. Huge sheets of calico with relevant diagrams helped the audience follow the lecture, a forerunner of presentation slides.

Discourses have collected all sorts of traditions, one of which is that the lecturer enters as the Theatre clock strikes the hour and talks until it chimes again. Lecturers also do not introduce themselves or introduce the lecture but instead plough straight into the subject.

From 1851 to 2001 most Discourses were recorded by the lecturer with a written overview published in the 'Proceedings of the Royal Institution', providing a valuable insight into the information that was presented.

SIR JAMES DEWAR LECTURING AT A FRIDAY EVENING DISCOURSE

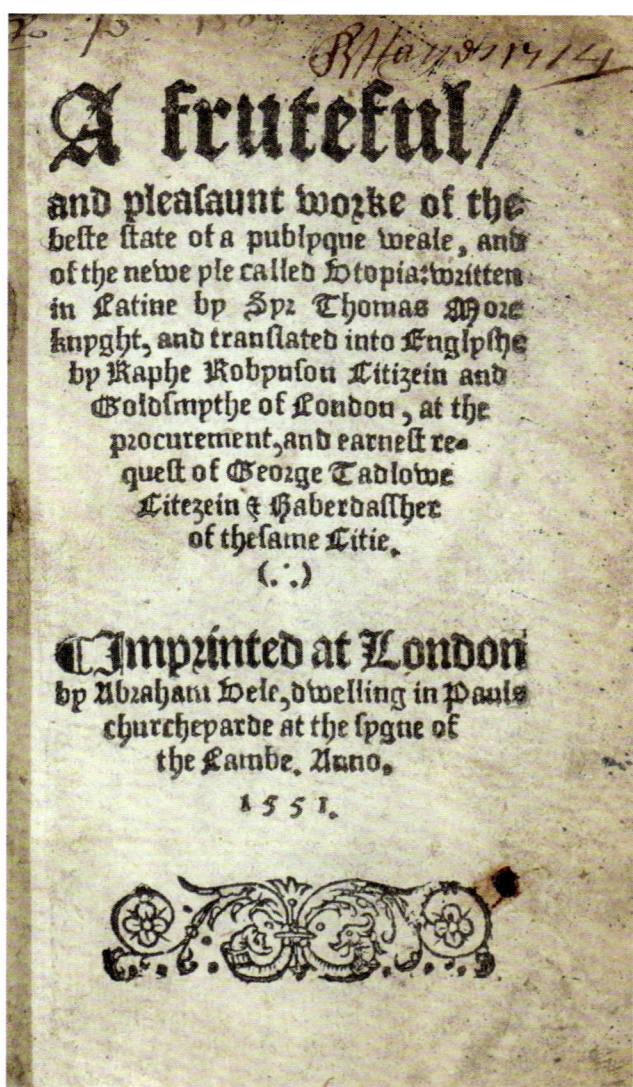

A fruteful,

and pleasaunt worke of the beste state of a publyque weale, and of the newe yle called Utopia: written in Latine by Syr Thomas More knyght, and translated into Englyshe by Raphe Robynson Citizein and Goldsmythe of London, at the procurement, and earnest request of George Tadlowe Citezein & Haberdassher of the same Citie.

Imprinted at London by Abraham Vele, dwelling in Pauls churcheyarde at the sygne of the Lambe. Anno.
1551.

THOMAS MORE (1478–1535)

Utopia, 1551

Paper, board, leather and printing ink, 140 mm × 85 mm × 20 mm
Gift to the Library of the Royal Institution by Michael Faraday, 1827
Ri RBMOR (FAR)

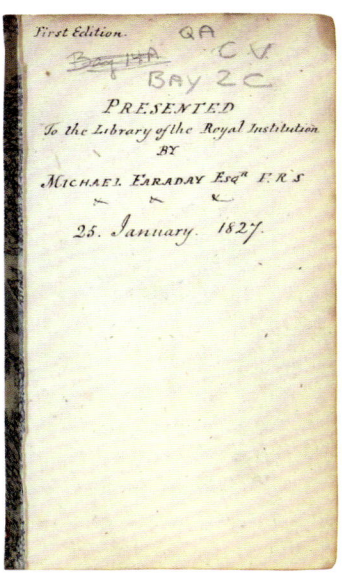

THIS SMALL VOLUME IS A FIRST EDITION OF THE English translation of Sir Thomas More's *Utopia*, transcribed by Raphe Robynson (also known as Ralph Robinson) in 1551.

Thomas More, a statesman and eventually Chancellor in the government of Henry VIII, wrote *Utopia* in 1516 in Latin. Utopia was written as a work of fiction, in which More depicts the imaginary island of Utopia, describing the religious, social, and political customs of its society and commentating on how the ideal state should be. Due to its lack of science, it is possibly not a book you would have expected to see in the library of a scientific institution. However, it was the breadth of its collection that made the Ri Library so famous throughout the 1800s and 1900s.

This copy of *Utopia* was originally owned by Michael Faraday, who donated it to the Ri Library in 1827. He gave several books to the Library during his time working at the Ri, to help increase the breadth of the offering. It is possible that some of the volumes, including this one, were even bound by Faraday himself prior to donation.

The Library was established in 1800, less than a year after the Ri moved into its home in Albemarle Street, London. Books were purchased in large quantities from this point on, in order for the Ri to have a well-stocked library for its members. The Library also took in books donated by its members to help swell these numbers.

The Library grew quickly and soon became an important place to visit, even featuring in Rudolph Ackermann's three-volume *Microcosm of London* (1808–10) as a place to see.

John Carr of York (1723–1807)

Grand Entrance, mid-1770s

Mahogany, iron and lead, various dimensions
21 Albemarle Street

The Ri entrance containing the impressive cantilevered grand staircase is the Ri's most complete surviving eighteenth-century interior, dating back to the mid-1770s and predating the founding of the Ri in 1799.

The staircase, with its elegant balustrade and mahogany handrail, was originally designed by the leading northern architect, John Carr. Over the years many documented changes have been made to the space, including the iron scroll brackets to support the stairs in the 1860s and the addition of the chandelier in the 1920s.

During the refurbishment in the 2000s, a conservator discovered that much of the eighteenth-century plasterwork had survived, allowing the original look and colour scheme to be established. It was therefore decided to restore the walls and balustrade to match the original design as far as possible. The words 'I. Simpson 1780' were found engraved on the underside of the handrail potentially identifying

an early craftsman, providing further evidence that the staircase dates to the late eighteenth century.

The Grand Entrance is dominated by the large statue of Michael Faraday, unveiled by the Prince of Wales in 1876. Started by John Henry Foley (1818–1874) and completed by Thomas Brock (1847–1922) after his death, its position in the entrance was supposed to be temporary, but it has in fact never been moved.

This space may feel intimidating, but it was originally the heart of the building, ceremoniously leading visitors to the Lecture Theatre one way and the Library the other, directing people to the Institution's two main areas for communication.

Sir Humphry Davy (1778–1829)

Experimental Gauze Safety Lamp, 1815

Also described as a miner's safety lamp and Davy lamp
Metal gauze and copper alloy, 189 mm × 89 mm
Made and used at the Royal Institution
RIAC 00205

This is the first completed miner's safety lamp, developed in the basement of the Ri in October to December 1815.

The Davy Lamp, as it is widely known, was developed by Sir Humphry Davy, assisted by Michael Faraday from October 1815. Davy was approached by a clergyman from the north-east of England who was losing many of his parishioners through explosions in the mines where they were working.

While deep underground extracting coal, miners only had candles to light their journey as well as their working environment. When heat from the flame met the pockets of methane gas that often existed around coal fissures, it would interact and cause an explosion.

After two months of experimenting and producing various prototypes, some with glass, Davy realised that encasing the candle with a simple section of metal gauze or mesh allowed the light to still pass through but absorbed the heat from the flame. It was therefore no longer hot enough to ignite the gas.

In early 1816 several copies of this lamp were tested in Hebburn Colliery near Newcastle, with instant success. The new safety lamp was immediately put into production and deaths were dramatically reduced. Davy's invention lives on through a modified version of his lamp, which is used today to transport the Olympic flame.

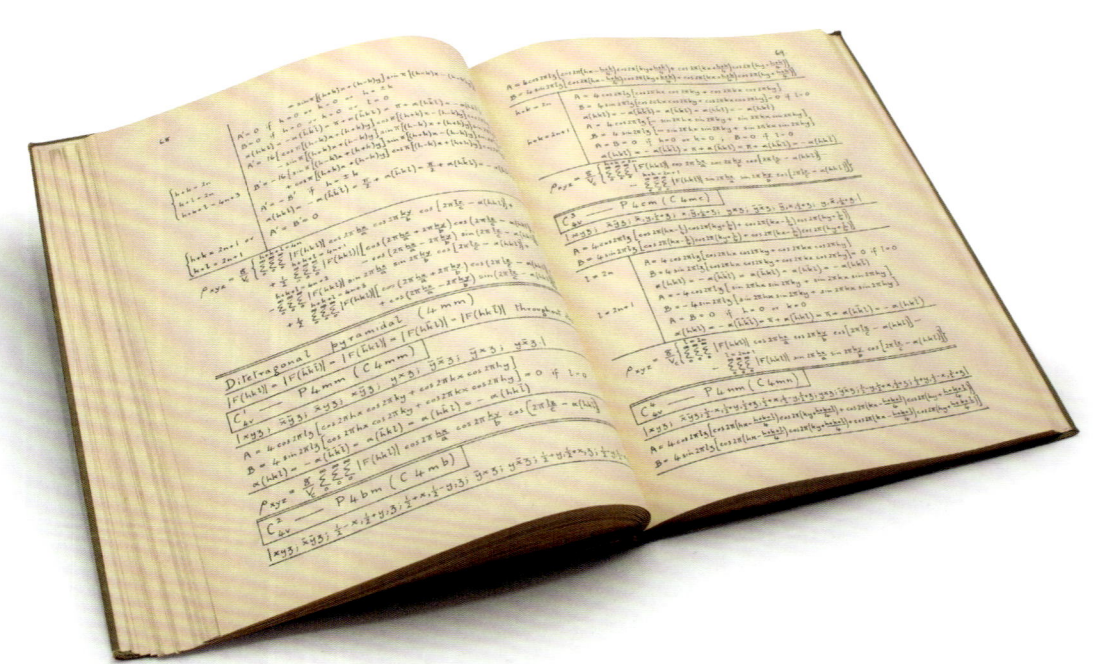

Kathleen Lonsdale (1903–1971)

Structure Factor Tables

Paper, printing ink, board, cloth binding, 287 mm × 225 mm × 15 mm
Published for the Royal Institution by G. Bell & Sons Ltd, London, 1936.
Ri RBLON (WHB)

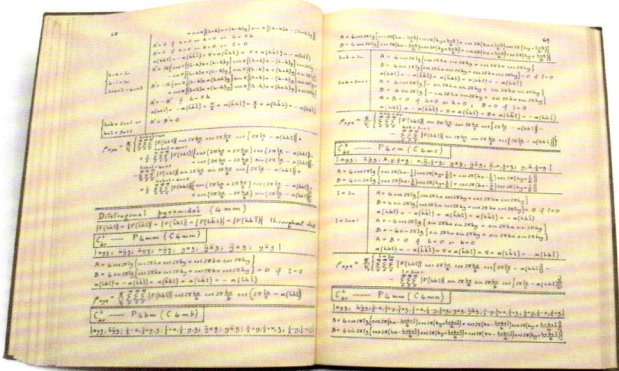

WRITTEN BY KATHLEEN LONSDALE, who worked in the laboratories of the Ri, this book was created to assist other researchers and help increase the scientific knowledge available to all. It was published in 1936 with the aim of improving access to fundamental equations in crystallography, speeding up research and collaboration.

Kathleen Lonsdale, née Yardley (1903–1971), was a crystallographer working in the research team of the Ri Director William Henry Bragg (1862–1942) in the 1920s and 1930s. Crystallography is the study of crystal structures, a branch of scientific research that had been pioneered by William Henry Bragg and his son William Lawrence (1890–1971) from 1913. Lonsdale's first major work was in the theory of space groups and to help determine when molecular symmetry occurs. She was the first to conclude and prove the structure of benzene, realising it was a linear hexagonal ring, and even calculated its precise dimensions.

Lonsdale neatly compiled this book by hand, filling it with tables and formulae she had noted over her years of research. This meticulous work was necessary for it to be reproduced in print without error. This was a unique and deeply personal publication, published by the Ri and made available to crystallographers and university departments in the UK. It helps illustrate that science is at most times collaborative and not just competitive.

Michael Faraday's Magnetic Laboratory

Lower ground floor, 21 Albemarle Street

IN 1819 THE SERVANTS' HALL, adjacent to the main Ri laboratory, was turned into a place for storing apparatus. By the 1850s, however, the space had become Michael Faraday's magnetic laboratory. A place where he could conduct his own research into electricity, magnetism and chemistry.

Much of the Faraday collection is still on display here, along with items used by such distinguished scientists as Sir Humphry Davy (1778–1829) and Sir Henry Cavendish (1731–1810). A large electromagnet, made from a link of a ship's anchor chain, sits proudly in the middle of the laboratory and was employed by Faraday in his work on diamagnetism. There are also less visible components to the room. Behind the rows of Cruickshank's 'troughs', or batteries, is a box of Faraday's early metal-alloy samples, and high on a shelf is his travelling watercolour set complete with unused pans of paint.

At the back of the room lies a service lift, a relic of when this was still a part of the kitchen of the original townhouse. Faraday had this space blocked up into cupboards, possibly to house chemical experiments on which he was working. On the cupboard door there is evidence of this usage as some of Faraday's wax seals remain, put in place to hold pieces of wire to prevent the opening of the doors until it was necessary.

John Tyndall (1820–1893)
Radiant Heat Apparatus, 1859–61

Various materials and multiple sizes
Made and used at the Royal Institution from the late 1850s
RIAC 00279_1-64

This collection of apparatus was used by John Tyndall to investigate the absorption and radiation of heat in gases, which resulted in the discovery of the principle behind the 'greenhouse effect'. Tyndall's experiments and results were published as a book 'Heat as a Mode of Motion' in 1863. Tyndall lectured on his experimentation concluding with his famous lecture '*On the Absorption and Radiation of Heat by Gases and Vapours, and on the Physical Connexion of Radiation, Absorption and Conduction – Bakerian Lecture, Published 1st January, read 7 February 1861*'.

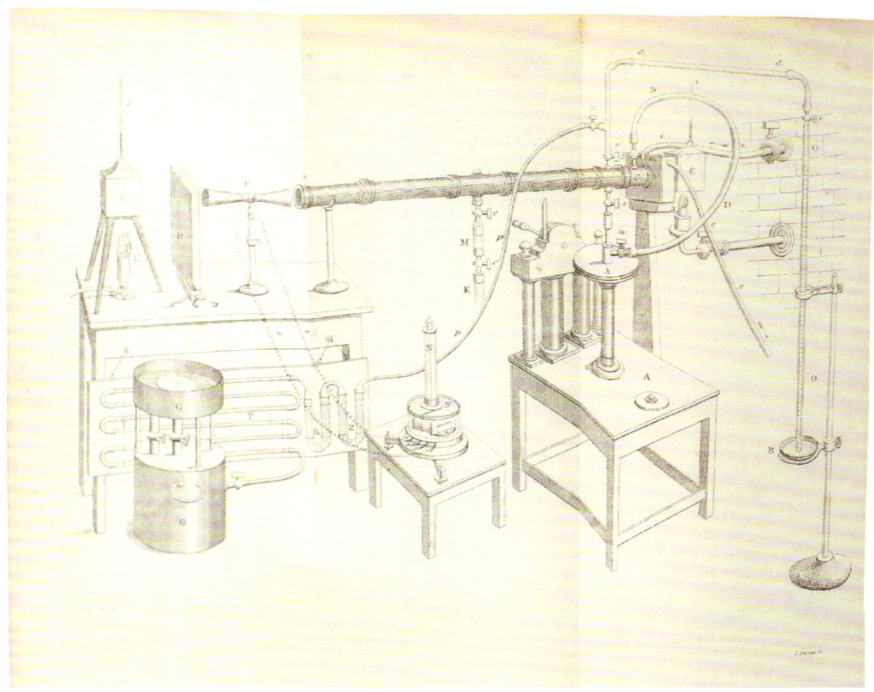

Tyndall wanted to know how different gases absorb heat from the Sun. He discovered that water vapour, carbon dioxide and ozone absorbed heat much more easily than the most common gases in our atmosphere, nitrogen and oxygen. He concluded that they have a greater influence on the temperature of the Earth's surface. He demonstrated that the strongest absorber was water vapour and therefore the most important of the gases. Without this gas, heat could be lost to outer space causing the Earth to cool too much to support human life. Therefore, a natural greenhouse is created through this absorbent layer of gas.

However, modern-day pollution affects these natural processes. Extra carbon dioxide and other gases lead to too much heat being trapped near the Earth's surface, causing the overall climate to become warmer.

Tyndall's equipment comprises heat tubes, Leslie cubes, rock-salt lenses and a thermopile that produced an electric current detected by a galvanometer. It was the swing of the needle on the latter that indicated the degree of heat absorbed by the gas in the tube.

Over the last decade we have come to understand that, at a similar time to Tyndall, Eunice Foote (1819–1888) was working in the USA and putting forward the same hypothesis about carbon dioxide and the climate, concluding that certain gases warmed when exposed to sunlight. She published her work in the *American Journal of Science* in 1857, but it seems that she had no knowledge of Tyndall's work nor him of hers.

Katherine Reynolds (c.1855–1932)

Photographic Collection, c.1890–1920s

Variable media and dimensions
Partly originated at the Royal Institution with some images donated by Reynolds's family in the 1930s
RIIC 0778 – 1077, plus additional loose images (approximately 200)

The Ri has a long history with the medium of photography, from the pre-development work of Humphry Davy to the advancements of William Henry Fox Talbot (1800–1877) that were announced at the Institution in 1839. The innovation was used by the Ri to illustrate lectures, capture experiments and raise the profile of the Professors. This connection with a developing medium was continued with the work of Katherine Reynolds (c.1855–1932), who took unprecedented images of the Ri and its functions and collections from the 1890s.

Sadly, little is known about Reynolds and her work. Her connection with the Ri is highly unusual, as she did not work for the Institution and was not a member in her own right. Her brother was the secretary of an engineering firm set up to develop an idea of Sir James Dewar, Director of the Davy Faraday Research Laboratory (DFRL) at the time. This relationship allowed her not only access to the Ri, but also it seems the freedom to roam and explore all areas of the building, including members spaces, public spaces, the Library (top, c.1890), the Director's flat and even the research laboratories. Clearly a keen amateur photographer, Reynolds produced hundreds of photographs, which are now in three bound volumes and two boxes of loose images. Her skill in photography is evident, but she also had a strong sense of detail, often documenting the time and day an image was taken and even the length of exposure and other technical information.

Her images help to illustrate the working of the Institution and how it has always displayed past apparatus and scientific developments for visitors to enjoy and learn from. By chance, she happened to capture significant events in her photographs, including the lecture bench used by Nikolai Tesla (1856–1943) in February 1892 (bottom), when he showed his developments in electricity to audiences of the Ri for the first time in the UK.

THOMAS PARKER (1772–1848)

Model of a Farm Gate, 1802

Various materials. Box: 169 mm × 259 mm × 201 mm; gate on stand: 150 mm × 181 mm × 178 mm.
Presented to the Royal Institution by Thomas N. Parker, February 1802
RIAC 00560

THIS OBJECT, COMPLETE WITH BOX, is a model of a working farm gate. It was presented to the Royal Institution in 1802 by its maker, Thomas Parker, and is one of the earliest models in the Ri's collection.

Parker was investigating how to prevent the destruction of crops by livestock. In the early 1800s the UK was about to enter a period of war. Cut off from Europe, the country needed to become self-sufficient, particularly in food production to feed workers. Agriculture, therefore, became an even more vital industry.

Parker's research revealed how crops were being destroyed when farmers and farmhands were not closing farm gates properly. Livestock were wandering through land where food crops were growing, trampling and eating much of the hard-earned produce.

In his treatise, *An Essay or Practical Inquiry Concerning the Hanging and Fastening of Gates and Wickets*, Parker described his investigations into how gates and wickets should be hung and fastened to ensure they may be 'perfectly in the same perpendicular line with each other, the gate will be at rest wherever it may be placed' and therefore be self-closing.

The boxed model, along with a pamphlet of Parker's investigations, was donated at the same time the model room was created at the Ri. The intention of this space was to use models to illustrate new scientific developments to the public.

The model consists of a circular wooden stand with three copper-alloy adjustable feet and a spirit level incorporated into its base. The stand has a square post at the centre, with various hinge points. The accompanying box has a key to lock it and the lid has three marks indicating where to place the stand in order to display it.

This model helps to demonstrate that scientific understanding can be found in everything, even in the simple principle of how to close a gate.

Postcard sent to Sir William Lawrence Bragg, 1913

Card, 140 mm × 90 mm
Donated by the Bragg family as part of the Sir William Lawrence Bragg Archive, Royal Institution
RI MS WLB/94/e/1

THIS POSTCARD IN THE archival collection of the Ri bears some of the most famous names in the world of physics at the start of the twentieth century. It was sent from Brussels by some of the participants of the second Solvay Conference to William Lawrence Bragg in October 1913, and can be described almost as a 'wish you were here' note. Written in German it reads, 'The most heartfelt congratulations to your wonderful scientific success and many greetings sent from Brussels from …' and is signed by 'Heinrich Rubens, A. Sommerfeld, M. Von Laue, W. Voigt, A. Einstein, M. Curie, E. Rutherford, F. Hasenöhrl, H. Lorentz, P. Weiss, Dr Broglie, P. Langevin, K. Onnes, J.J Thomson, W. Nernst, E. Warburg and Martin Knudsen'.

The first international conference had taken place in Brussels in 1911, at which the leading physicists of the day discussed the 'theory of radiation'. The success of this meeting encouraged its founder, Ernest Solvay, to donate a huge sum of money to form an International Physical Institute.

Under the presidency of Professor Lorentz, the second conference confined its discussion to the structure of the atom and the structure of crystals. Given the work that both William Henry and William Lawrence Bragg had done on the development of X-ray crystallography in 1912/13, it seems obvious that at least one of them would be invited to attend.

While William Henry Bragg was able to join the discussion in Brussels, it appears his son was yet to be acknowledged.

This postcard shows there was a willingness to communicate and discuss scientific theory and developments not only between scientists and the public, but also among the scientists themselves, who met to work through ideas in order to advance them.

POSTCARD SENT TO WILLIAM LAWRENCE BRAGG | **55**

Sir James Dewar (1842–1923)

Early Isolation Vessel, 1892

Also known as a Dewar Flask, Vacuum Flask or Thermos Flask
Glass, wood and shellac, 254 mm × 128 mm
Made and used at the Royal Institution
RIAC 00420

This is the earliest form of a double-walled isolation vessel or vacuum flask, made by Sir James Dewar in 1892, and consists of two round-base, glass inner vessels sitting in an outer cylindrical jar with a sealed wooden lid. Vacuum flasks, or Dewar Flasks, are characterised by a vacuum between double walls of glass, allowing the contents to stay very cold or very hot. Dewar's invention is probably more commonly known globally today as the Thermos™ flask.

At the time of making, Dewar was Director of the Davy Faraday Research Laboratory (DFRL) at the Royal Institution, investigating cryogenics, or the science of extreme cold. He realised that he needed a way to stop the very cold liquids with which he was working from evaporating as they heated up. Experimenting with various methods, including insulated boxes filled with straw, he developed the vacuum flask.

Dewar designed the double-walled glass flask with a vacuum between each layer, realising that heat could not be transferred through the vacuum. His invention took shape while he was working on the magnetic properties of liquid oxygen. To improve the flask, he eventually added silvering to the outside surface, allowing it to work even better at keeping its contents cold for longer.

Dewar used the flask to liquefy hydrogen, which was the coldest substance to have been produced by that date. To liquefy hydrogen, he had to create temperatures of -252.5°C or 20°C above absolute zero, the lowest recorded at the time.

MICHAEL FARADAY (1791–1867)

Magneto-Optical Effect Apparatus, 1845

Iron, calico, metal wire, glass and wood, 385 mm × 263 mm × 167 mm
Made and used at the Royal Institution
RIAC 0003

THIS APPARATUS WAS USED BY Michael Faraday to discover and demonstrate the magneto-optical effect, showing that light and glass are affected by magnetism. The phenomenon is now known as the 'Faraday Effect'. The apparatus is made up of a combination of items from other experiments, including a large electromagnet bound in heavy wire that was created in the 1830s and a block of pure glass that Faraday had made in the late 1820s, which sits on top.

Faraday was trying to prove that all matter was magnetic and that there was a fundamental link between electricity, magnetism and light. From 1845 in the basement of the Ri, he undertook a series of experiments using a dense piece of glass and an electromagnet. Faraday found that when light was transmitted through the glass along the direction of an external magnetic field, its polarisation angle was rotated. This is the first time that the magneto-optical effect, was observed providing proof that light is an electromagnetic wave in essence.

This experiment helped to provide experimental evidence for Faraday's belief in the unity of the fundamental forces of nature. It illustrated that electricity and magnetism could affect how light behaves.

Faraday spent much of the rest of the 1840s exploring the magneto-optical effect, obtaining the same results with more and more compact pieces of apparatus.

Sir George Porter (1920–2002)
Ruby Laser, c.1966

Metal, glass, aluminium oxide and synthetic ruby, 329 mm × 597 mm × 200 mm
Made and used at the Royal Institution
RIAC 0069

THIS IS ONE OF THE FIRST RUBY LASERS ever used in the UK. Constructed at the Royal Institution *c.*1966, it was employed by Ri Director Sir George Porter's research team to capture chemical reactions and study chemical molecules that last for only a fraction of a second. Porter (1920–2002) was Director of the Ri from 1966 to 1986, as well as holding the positions of Professor of Chemistry and Director of the Davy Faraday Research Laboratory (DFRL).

In 1960, Theodore Maiman, an employee of the Hughes Research Lab in California, successfully constructed the world's first working laser, made of synthetic ruby. The laser consists of a ruby tube in a hinged cylindrical casing, which is fixed horizontally to a metal stand. Using the laser, Porter was able to demonstrate the technique of flash photolysis. This process of using flashes of light to enable chemical reactions to be examined was first developed in 1949 and it was for this work that Porter, along with Manfred Eigen and Ronald George Wreyford Norrish received the 1967 Nobel Prize in Chemistry.

Porter's research team at the Ri captured chemical reactions lasting less than a nanosecond, which is one billionth of a second. With increasingly sophisticated lasers, the team was finally able to work with picoseconds, or a thousandth of a billionth of a second.

Without this technology and area of research we would never have known how chlorofluorocarbons (CFCs) break down the ozone layer, or why fabric fades in sunlight.

Terence Cuneo (1907–1996)

Sir William Lawrence Bragg Giving the 1961 Christmas Lectures on Electrostatics in the Ri Lecture Theatre, 1962

Oil on canvas, 1390 mm × 1920 mm × 80 mm
Commissioned by the Royal Institution in 1961 from the artist; completed in 1962
RIIC 0621

This painting by Terence Cuneo is a faithful depiction of Sir William Lawrence Bragg's Christmas Lectures undertaken at the Ri in 1961. Bragg was the Ri's Director, Director of the Davy Faraday Research Laboratory (DFRL) and Professor of Chemistry at this time. Also depicted is William 'Bill' Coates (1919–1993), a long-serving lecture and laboratory assistant, a legendary figure at the Ri. He had an invaluable way of being able to devise demonstrations that illustrated scientific principles. The audience is made up of schoolchildren, with Lord Brabazon of Tara (1884–1964) in the President's chair.

Scattered across the bench are various pieces of electrostatic equipment, including a Van de Graaff generator and a Wimshurst machine being operated by Coates. Coates often used the most basic items to make his own apparatus: for example, the two large red balls with cotton attached seen in the painting were employed to show the effects of static electricity. These balls are from the valves found in toilet cisterns.

The painting helps to highlight Bragg's enthusiasm for giving the Christmas Lectures. He was particularly passionate about lecturing to children and engaging audiences with science and discovery. While in charge at the Ri he started the schools programme and assisted in getting the Christmas Lectures televised as complete series, with the first one broadcast in 1966.

Terence Cuneo enjoyed an illustrious career as an artist throughout the twentieth century. His prestigious commissions included works for the War Artists' Advisory Committee and portraits of Elizabeth II – he was the official artist of her Coronation. One of Cuneo's amusing motifs was to include a small mouse somewhere in the picture. Can you spot our extra audience member?

Model of Lysozyme, c.1965

Perspex, felt and wood, 625 mm × 403 mm × 356 mm
Originally constructed at the Royal Institution Workshop, 1965
RIAC 00701

THIS IS A MOLECULAR MODEL, made of painted wire supported on a Perspex pole and fixed on a felt-covered wooden base. It was produced in the 1960s to illustrate the structure of lysozyme, which is invisible to the naked eye.

Lysozyme is an enzyme found in many biological organisms. In humans it exists in tears, saliva and mucus. First discovered by Alexander Fleming (1881–1955) in 1922 when mucus from his nose accidentally mixed with a sample of cultured bacteria, causing it to dissolve. He realised that something within the substance must be antibiotic in nature and named it Lysozyme. Without knowing the structure of this substance, however, scientists could not determine fully how it worked or reacted.

Within the Davy Faraday Research Laboratory (DFRL) at the Ri, a team led by the then Director, Sir William Lawrence Bragg, and which included David Chilton Phillips (1924–1999) and Louise Johnson (1940–2012), used a technique called X-ray crystallography to determine the enzyme's structure.

When a beam of X-rays was shone through a sample of lysozyme, the X-rays bounced off the atoms to create a distinctive pattern. An early computer was used to analyse the pattern and re-create the 3D arrangement of atoms within the molecule, similar to the way we use X-rays to see inside the human body today. This pattern is represented by this model. With the completion of the work in 1965, lysozyme became the first enzyme ever to have its structure identified.

These models were important to X-ray crystallographers, allowing them to visualise the structure of a substance or material in order to better understand them and help determine their functions and properties for future use. This paved the way for the creation of modern life-saving drugs through the manipulation of molecular structures.

Michael Faraday (1791–1867)

Colloidal Samples and Microscope Slides, 1850s

Various dimensions and materials
Made and used at the Royal Institution from the 1850s
RIAC 00104_a-e (Colloidal Bottles); RIAC 00107_a-f (Colloidal Bottles); RIAC 00124_a-h (Colloidal Slides Boxes)

In 1853 at a lecture at the Royal Institution, Michael Faraday demonstrated a fascinating property of metallic gold: in contrast to all other metals, 'Gold has been beaten into leaves so fine as to become partially transparent, not in consequence of any cracks, holes or fissures, but by the shining of light through its substance'.

Besides transparency, the gold leaf possessed another remarkable property. When light was passed through it, it appeared green, not yellowish gold. The 'shining of light through its substance', as Faraday noted, had changed the light.

Faraday continued his research and in 1856 discovered the first metallic colloid of gold. 'Faraday's Gold', as it became known, was a mixture of two or more solids, liquids, or gases. Faraday was particularly interested in the dispersion of the very fine gold particles he could suspend throughout the liquid. This type of preparation is known as a colloidal suspension or, as Faraday named it, a gold 'sol'. He conducted hundreds of individual experiments in this area of research and prepared over a thousand specimens. The Ri holds more than 700 microscope slides and 15 bottles of colloidal solutions. Interestingly, many of these colloids are still optically active. If we follow Faraday and shine a modern laser pointer through these bottles a cone of light is produced, demonstrating what is known as the Faraday-Tyndall effect. It is still a mystery why Faraday's colloids continue to work after more than 150 years since their creation.

Warren De La Rue (1815–1889)

Glass Plate Photographs of the Moon and Sun, 1858–61

Various materials and dimensions
Gift in the will of Warren De La Rue
RIDM 0226_1-10 and RIDM 0227_1-14

These boxes of glass plates were Warren De La Rue's personal collection of photographs, which he took through the 1850s and 1860s. The images depict the Moon and even capture a full eclipse of the Sun.

De La Rue was an active member of the Ri, not only lecturing but also undertaking administrative duties. He was Secretary of the Institution from 1879 to 1882.

He began experimenting with lunar photography, using wet-collodion glass negatives and a telescope of his own design. He was among the earliest astronomer-photographers and worked at Kew Observatory (also known as the King's Observatory at Richmond) as well as travelling internationally to take images. He took photographs of the Moon and produced a series of spectacular stereoscopic plates that give the appearance of depth and luminosity. These images are beautiful representations of an experiment, the desire to capture a fleeting or hard-to-reach or see event and share it with the wider world.

Most of the Moon images were taken between 1858 and 1861. During an expedition to Spain to view a total eclipse in July 1860, De La Rue produced photographs of the event and returned to England to lecture at the Ri and Royal Society about his investigations.

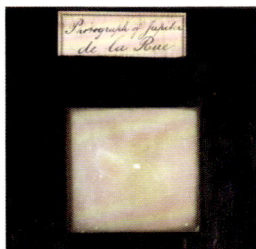

MICHAEL FARADAY (1791–1867)

First Sample of Benzene, 1825

Glass, cork and benzene liquid, 146 mm × 44 mm
Made and used at the Royal Institution in 1825
RIAC 00050

BENZENE IS AN IMPORTANT ORGANIC COMPOUND known as a hydrocarbon. Colourless and flammable, with a sweet odour, it is a highly dangerous but incredibly important substance. It is a widely used industrial chemical that can be found in crude oil, plastics, resins, dyes, detergents, pesticides, drugs and synthetic fibres.

This small vial of liquid is the first ever sample of benzene, isolated by Michael Faraday in 1825. He extracted a liquid from cylinders of compressed illuminating gas that had been collected from whale oil by the process of pyrolysis. After making the sample, Faraday announced the existence of 'bicarburet of hydrogen', later named benzene and even etched this information onto the side of the vial. He did not pursue his discovery however, instead leaving it to others to determine if this new liquid had any usefulness.

In the 1830s Faraday agreed to a request from a German chemist, Eilhard Mitscherlich (1874–1956), to change the name of the compound to 'benzene'. Mitscherlich had been working on distilling calcium salts and benzoic acid, producing a product identical to Faraday's.

Its usefulness can be put down to its unique ring structure of six carbon atoms combined with only six hydrogen. The work to determine the structure of benzene was undertaken at the Ri and Leeds University by Kathleen Lonsdale, a researcher in Sir William Henry Bragg's team. Lonsdale's breakthrough research was completed in 1928, just over 100 years after Faraday originally discovered the chemical. She demonstrated that the benzene ring structure could not be anything but a flat hexagon, and provided accurate distances for all carbon-carbon bonds in the molecule. Lonsdale published a short summary of her findings in *Nature* on 24 November 1928.

Sir Humphry Davy (1778–1829)

Parabolic Mirrors, *c.*1804

Copper, tin and iron, 210 mm × 610 mm
Made and used at the Ri from the early 1800s
RIDM 0313

These objects are parabolic mirrors. Highly reflective and concave in shape, they were used by Sir Humphry Davy at the Royal Institution to demonstrate that heat was a form of motion, challenging the existing theory that it was a fluid and building on the work of Sir Benjamin Thompson (1753–1814), Count Rumford.

Davy suspended one mirror from the ceiling of the Lecture Theatre and hung a small bag of gunpowder from the hook in the middle. The second mirror was positioned on the floor directly beneath it. A red-hot cannonball was then placed in the centre of this second mirror. Heat from the cannonball was reflected from one mirror to the other, with the second mirror focusing the heat on the gunpowder above, causing a dramatic explosion. Davy showed that heat travels in waves, like light, and can be reflected as radiant heat. He and Count Rumford continued this work, discovering that shiny surfaces reflect more heat than dull, rough ones, and that an object can radiate more heat in a vacuum.

It was spectacular demonstrations like these that drew crowds to the Ri Lectures. On some lecture evenings there were so many carriages dropping off audience members that a one-way system had to be instigated, making Albemarle Street the first one-way street in London.

Cranium Coil and Magnetic Agent, 2022

Metal, copper wire, iron oxide, insulation material and plastic, 110 mm × 270 mm × 190 mm
Donated by Professor Quentin Pankhurst, Professor of Physics Director Institute of Biomedical Engineering, University College London, 2023
RIAC 03700

This bright yellow ring is one of the Ri's most recent acquisitions. It came from a team of researchers at Resonant Circuits Limited who work in the Ri's laboratories and workshops, continuing the 200-year-plus tradition of scientific research on the premises. Led by Professor Quentin Pankhurst of University College London, the researchers are using the phenomenon of electromagnetic radiation – first characterised by Michael Faraday's work at the Institution – to find and treat cancer.

The yellow ring is designed to rest on the top of a patient's head like a crown. Inside, copper wire is coiled tightly around a curved metal rod. When the 'crown' is connected to the electricity supply, the electricity flowing through the coils induces a time-varying magnetic field that extends around the patient's head.

Stored in the vial are iron oxide nanoparticles, which are very similar to the gold nanoparticles first discovered by Faraday. They are injected into the patient's body and disperse inside the cancer tumour. When the time-varying magnetic field is applied, the tiny nanoparticles heat up – and by heating up, they damage the cancer tumour. Clinical trials are currently underway to test whether this promising application could be scaled up to become a mainstream therapy.

The modern objects in the Ri collections are just as valuable as the older ones because they remind us that science does not stand still; fresh discoveries and applications are always just around the corner. This bright yellow ring and the vial of nanoparticles also symbolise the Ri's distinctive nature as a home for cutting-edge current science as well as a place in which to showcase science from the Institution's past.

Sir William Lawrence Bragg (1890–1971)

'Bragg's Law' Notebook, 1912

Paper, board, ink, card and cloth binding, 230 mm × 180 mm × 290 mm
Donated by the Bragg family along with scientific papers of W.L. Bragg
RI MS WLB/86

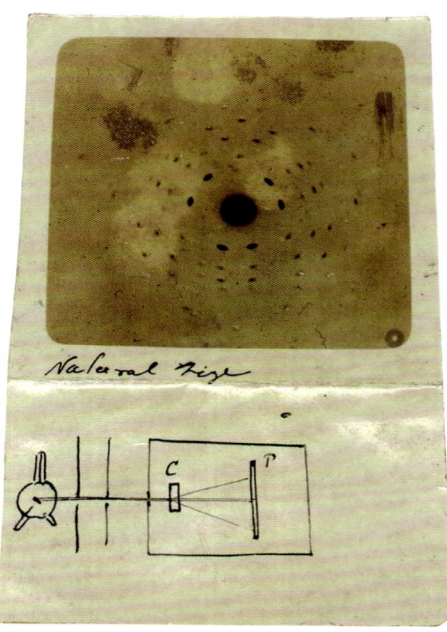

This is the laboratory notebook of William Lawrence Bragg from 1912, when he was investigating the structure of crystals.

In the summer of 1912 a German physicist, Max von Laue (1879–1960), working in Munich, discovered the diffraction of crystals by X-rays. In England, William Henry Bragg, Professor of Physics at the University of Leeds, investigated Laue's work. Conferring with his son, William Lawrence, they set about proving that the phenomenon can be simply understood in terms of the reflection of X-rays by planes of atoms in the crystal. This became 'Bragg's Law' and its development, along with the specialist spectrometer used to analyse the crystal structure, helped secure this father-and-son team the Nobel Prize in Physics in 1915. Aged 25 at the time, William Lawrence Bragg is still the youngest ever scientific Nobel Prize winner.

This work opened the way to the detailed study of X-rays and, at the same time, began the X-ray analysis of crystal structures that has since revealed the arrangement of atoms in all kinds of substances, from chemical elements to viruses.

The wave length of the reflected light will be.

$2 \cdot d \cos\theta.$ $d =$ perp distance of planes apart.

$= 2 \cdot \left(\dfrac{d}{\cos\theta}\right) \cos^2\theta.$

$= 2 \cdot \dfrac{a}{p_1} \cdot \gamma^2$

where γ is the direction cosine of the reflecting plane.

The plane makes intercepts.

$p_s \quad q_n \quad q_s.$

$\gamma = \dfrac{\frac{1}{q_s}}{\sqrt{\frac{1}{p_s^2}+\frac{1}{q_n^2}+\frac{1}{q_s^2}}} = \dfrac{p_n}{\sqrt{p^2 n^2 + p^2 s^2 + q^2 n^2}}$

Wavelength $= \dfrac{2 \cdot a \cdot p_1^{22}}{p^2 n^2 + p^2 s^2 + q^2 n^2} \cdot \dfrac{1}{L.C.M.(p_1)}$

$\dfrac{\lambda}{a} = \dfrac{2 p_1^{22}}{p^2 n^2 + p^2 s^2 + q^2 n^2} \cdot \dfrac{1}{L.C.M.(p_1)}$

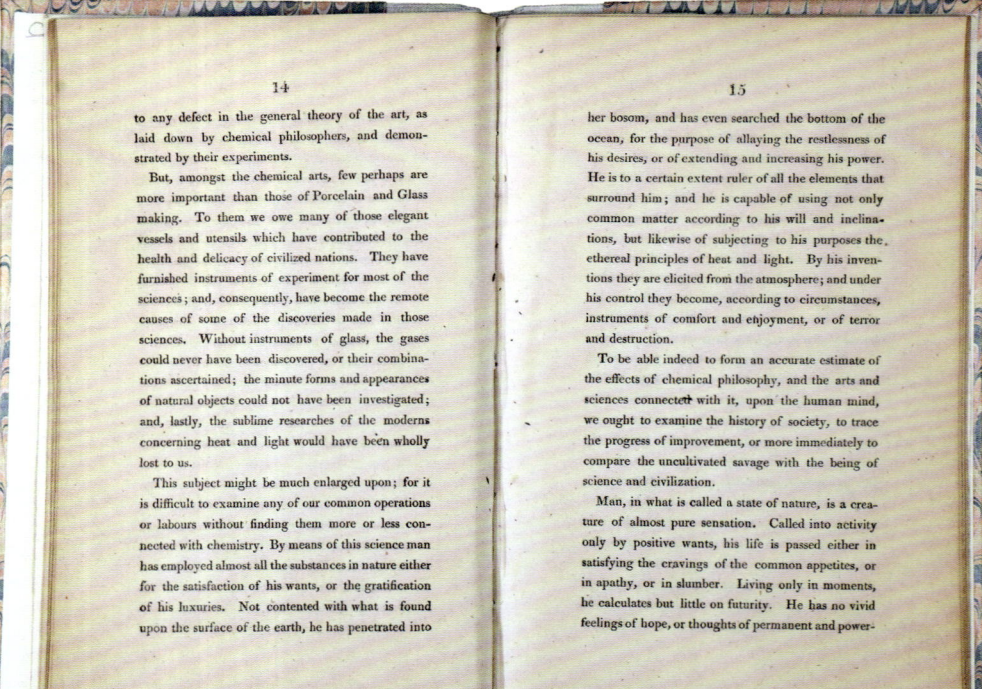

to any defect in the general theory of the art, as laid down by chemical philosophers, and demonstrated by their experiments.

But, amongst the chemical arts, few perhaps are more important than those of Porcelain and Glass making. To them we owe many of those elegant vessels and utensils which have contributed to the health and delicacy of civilized nations. They have furnished instruments of experiment for most of the sciences; and, consequently, have become the remote causes of some of the discoveries made in those sciences. Without instruments of glass, the gases could never have been discovered, or their combinations ascertained; the minute forms and appearances of natural objects could not have been investigated; and, lastly, the sublime researches of the moderns concerning heat and light would have been wholly lost to us.

This subject might be much enlarged upon; for it is difficult to examine any of our common operations or labours without finding them more or less connected with chemistry. By means of this science man has employed almost all the substances in nature either for the satisfaction of his wants, or the gratification of his luxuries. Not contented with what is found upon the surface of the earth, he has penetrated into her bosom, and has even searched the bottom of the ocean, for the purpose of allaying the restlessness of his desires, or of extending and increasing his power. He is to a certain extent ruler of all the elements that surround him; and he is capable of using not only common matter according to his will and inclinations, but likewise of subjecting to his purposes the ethereal principles of heat and light. By his inventions they are elicited from the atmosphere; and under his control they become, according to circumstances, instruments of comfort and enjoyment, or of terror and destruction.

To be able indeed to form an accurate estimate of the effects of chemical philosophy, and the arts and sciences connected with it, upon the human mind, we ought to examine the history of society, to trace the progress of improvement, or more immediately to compare the uncultivated savage with the being of science and civilization.

Man, in what is called a state of nature, is a creature of almost pure sensation. Called into activity only by positive wants, his life is passed either in satisfying the cravings of the common appetites, or in apathy, or in slumber. Living only in moments, he calculates but little on futurity. He has no vivid feelings of hope, or thoughts of permanent and power-

Sir Humphry Davy (1778–1829)

Published Royal Institution Lecture by Humphry Davy, 1802

Paper, printing ink and board, 210 mm × 135 mm × 5 mm
Copy obtained by the Royal Institution Library, 1802
RI RBDAV

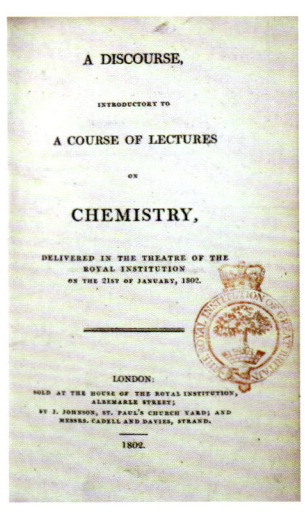

While the Ri holds Davy's original handwritten lecture notes, these are quite difficult to read, and this printed book, no bigger than a pamphlet, allows for a better understanding of what was conveyed in the lecture. It highlights the work the Ri undertook during its early years to disseminate more widely, and beyond the initial audience, the teachings that occurred within the Lecture Theatre, with printed copies being made available for purchase shortly after the lecture took place.

Davy was employed at the Ri from 1801, later becoming Director of the Laboratory and Professor of Chemistry. He helped to establish the Ri's reputation for giving excellent lectures, conveying his scientific research into chemical reactions through spectacular demonstrations. It was therefore only natural for the Ri to publish the content of these lectures for others to continue to use in home study.

This book contains a transcript of Davy's 'Introductory Course of Lectures on Chemistry' delivered in the Theatre of the Royal Institution on 21 January 1802. It is the culmination of Davy's chemical research, in which he argued that the power of chemistry was key to understanding life and the natural world. Mary Shelley (1797–1851) took inspiration from this viewpoint, basing her character of Professor Waldman in her novel *Frankenstein* (1818) directly on Davy. It is Waldman who inspires Frankenstein to discover the secret of life.

This book has been made possible thanks to the kindness and generosity of the Faraday Foundation, Wayne Kamitaki, and Carla & Steven Jacobs.

First published in 2024 by
Scala Arts & Heritage Publishers Ltd
43 Great Ormond Street
London WC1N 3HZ
www.scalapublishers.com
An imprint of B.T. Batsford Holdings Ltd.
In association with The Royal Institution of Great Britain, 2024.

10 9 8 7 6 5 4 3 2 1
ISBN 978-1-78551-530-9

This edition © B.T. Batsford Ltd., 2024. Text and images © in association with The Royal Institution of Great Britain, 2024. All rights reserved. No part of this book may be reproduced, stored in a retrieval system, or transmitted in any form or by any means electronic, mechanical, photocopying, recording or otherwise, without the written permission of Scala Arts & Heritage Publishers Ltd and The Royal Institution.

Editors: Linda Schofield and Claire Young
Design: Joe Ewart
Printed and bound in Turkey by Elma Basim

Director's Choice is a registered trademark of Scala Arts & Heritage Publishers Ltd.

FRONT COVER:
Terence Cuneo, *Sir William Lawrence Bragg Giving the 1961 Christmas Lectures on Electrostatics in the Ri Lecture Theatre*, 1962. See pp. 62–63.

FRONTISPIECE:
John Carr of York, Grand Entrance, mid-1770s. See pp. 40–41.

BACK COVER:
Model of Lysozyme, *c.*1965. See pp. 64–65.

AUTHOR ACKNOWLEDGEMENTS
I would like to extend thanks to colleagues at the Ri who have contributed to this publication, particularly Charlotte New, whose extensive knowledge of the Ri's heritage, collection and archive has been invaluable.

Unless noted below, all photographs are by Paul Wilkinson and © The Royal Institution: p. 30, Steven Franklin; p. 41, Matt Chung.